AF898703

discovermore

Earth's Processes

What Are Weather and Climate?

Doug Bradley

Britannica Educational Publishing
IN ASSOCIATION WITH
THE ROSEN PUBLISHING GROUP

Published in 2027 by Britannica Educational Publishing (a trademark of Encyclopædia Britannica, Inc.) in association with The Rosen Publishing Group, Inc.
2544 Clinton Street, Buffalo, NY 14224

Distributed exclusively by Rosen Publishing.
To see additional Britannica Educational Publishing titles, go to rosenpublishing.com.

Editor: Deanna Lepovich
Book Design: Greg Roza

Photo Credits: Cover; (series background) Dai Yim/Shutterstock.com; cover ARTpok/Shutterstock.com; p. 5 (top) TORWAISTUDIO/Shutterstock.com; p. 5 (bottom) Alena Mostovich/Shutterstock.com; p. 6 Olga_Kuzmina/Shutterstock.com; p. 7 Mamunur Rashid/Shutterstock.com; p. 9 (top) kazutaka.Japan/Shutterstock.com; p. 9 (bottom) Berit Kessler/Shutterstock.com; pp. 10, 21 (bottom) VectorMine/Shutterstock.com; p. 11 Phototee/Shutterstock.com; p. 12 Dmitry Molchanov/Shutterstock.com; p. 13 Lukas Jonaitis/Shutterstock.com; p. 14 Romberi/Shutterstock.com; p. 15 Gorodenkoff/Shutterstock.com; p. 16 Patrik Slezak/Shutterstock.com; p. 17 Andrei Armiagov/Shutterstock.com; p. 19 (top) Everett Collection/Shutterstock.com; p. 19 (bottom) manfredrichter/Shutterstock.com; p. 21 (top) Mozgova/Shutterstock.com; p. 22 grayjay/Shutterstock.com; p. 23 Chris Christophersen/Shutterstock.com; p. 25 (top) jantsarik/Shutterstock.com; p. 25 (bottom) Digi.Pixel/Shutterstock.com; p. 26 Aleksander Markov/Shutterstock.com; p. 27 Bagus upc/Shutterstock.com; p. 28 Svetlana Mahovskaya/Shutterstock.com; p. 29 Fahrenheit small/Wikimedia Commons.

Cataloging-in-Publication Data

Names: Bradley, Doug, 1971-.
Title: What are weather and climate? / Doug Bradley.
Description: New York : Britannica Educational Publishing, in association with Rosen Educational Services, 2027. | Series: Discover more: Earth's processes | Includes glossary and index.
Identifiers: ISBN 9781641906050 (library bound) | ISBN 9781641906043 (pbk) | ISBN 9781641906067 (ebook)
Subjects: LCSH: Weather--Juvenile literature. | Meteorology--Juvenile literature. | Climatic changes--Juvenile literature. | Climatology--Juvenile literature.
Classification: LCC QC981.3 B656 2027 | DDC 551.6--dc23

Printed in China

Contents

What Should I Wear Today?

The weather is important to people. It affects our comfort, food supply, and even safety. Weather is the daily state of the atmosphere, or air, in any given place. There are many elements that combine to create weather. These elements vary from region to region. They include temperature, wind, humidity, precipitation, and atmospheric pressure.

Climate is the weather found in a certain place over a long period of time. An area's climate determines what kinds of plants can grow and what kinds of animals can survive there. People use information about climates to decide which crops to plant and prepare for natural disasters. The climate greatly affects how people travel and where they choose to go.

On rainy days, people often wear raincoats or use umbrellas to stay dry.

compare and contrast

How are weather and climate similar? In what ways are they different?

When it's sunny, it's a good idea to wear a hat and sunglasses to protect our skin and eyes from the sun's powerful rays.

Where Does Wind Come From?

Temperature and wind are two basic elements of weather. Temperature is how warm or cold it is outside. Temperatures often change with the time of day or year. Wind is the movement of air near Earth's surface. Wind can be a light breeze or a wild gale. The most powerful wind happens during storms like tornadoes, cyclones, typhoons, and hurricanes.

We use wind in many ways.
We couldn't fly kites without wind!

Monsoon season in Bangladesh lasts from June to September. Monsoons can result in massive flooding.

Changes in the temperature of air, land, and water cause wind. When air flows over a warm surface, it heats up and rises. This leaves room for cooler air to flow in. The flowing air is wind.

Winds greatly affect the weather. They bring cool air into warm areas. They also can bring rain, snow, or even dust and sand. For example, in southern Asia, winds known as monsoons bring rain during the summer when cool, moist air from the ocean moves over the warm land.

Consider This

How windy does it get where you live?
What are the strongest winds you've ever experienced?

Moisture in the Air

A cloud is made up of millions of tiny water droplets or ice crystals floating together in the air. On a sunny day a wispy cloud can be a beautiful sight, but at other times a cloud can be a sign of storms to come.

The air always contains water vapor—water in the form of gas—which is invisible. The amount of moisture in the air is known as humidity.

The amount of water vapor that air can hold depends on the air's temperature. The cooler the air, the less water it can hold. When air cools, some of the water vapor condenses, or forms visible water droplets. The droplets form around tiny particles in the air, such as dust or sea salt. Up in the sky, it forms clouds. Near the ground, the condensed water vapor becomes fog.

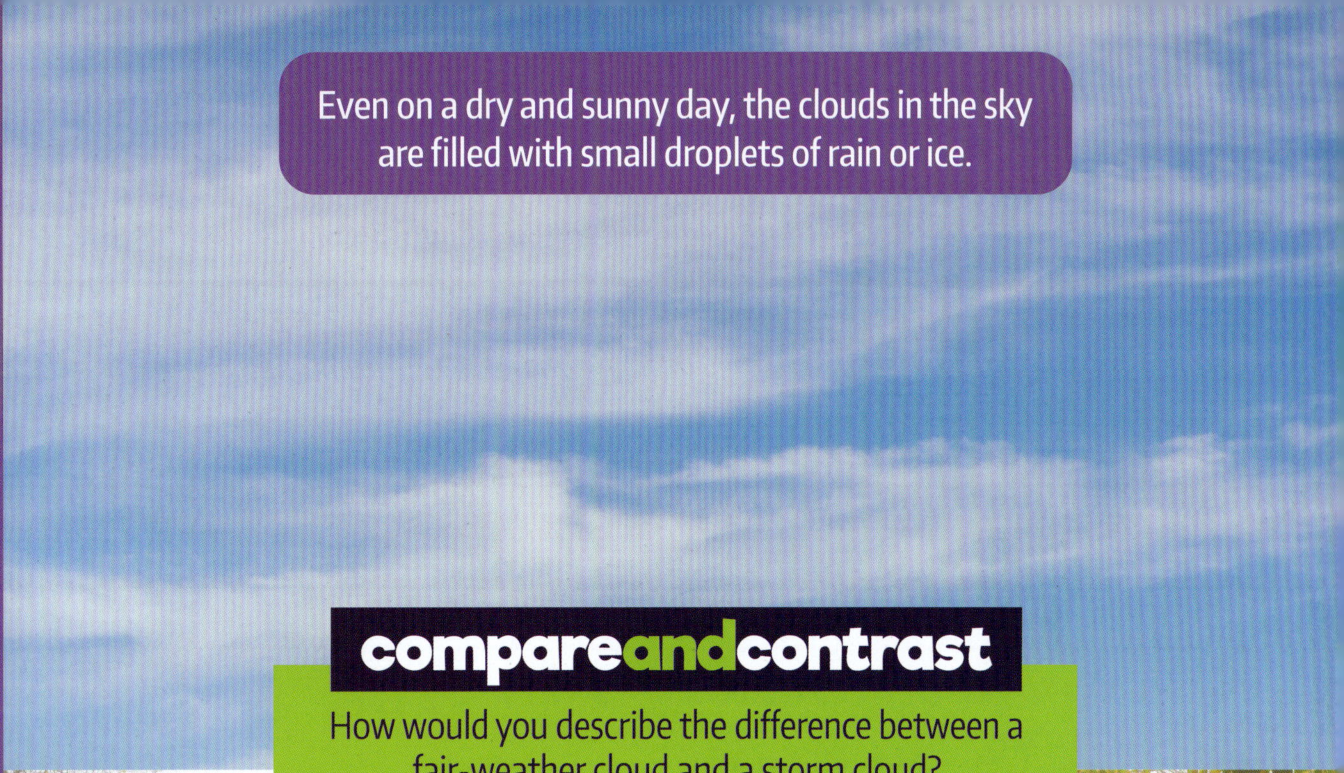

Even on a dry and sunny day, the clouds in the sky are filled with small droplets of rain or ice.

compare and contrast

How would you describe the difference between a fair-weather cloud and a storm cloud?

Fog is made up of tiny particles of water floating in the air. It's much like a cloud that forms near the ground.

Precipitation

Precipitation is a major part of Earth's endless **water cycle**. Precipitation comes in many forms, including rain, hail, sleet, and snow. The form precipitation takes depends on other weather conditions, such as temperature.

This diagram shows how Earth's water cycle works. Earth's water is constantly moving through this cycle.

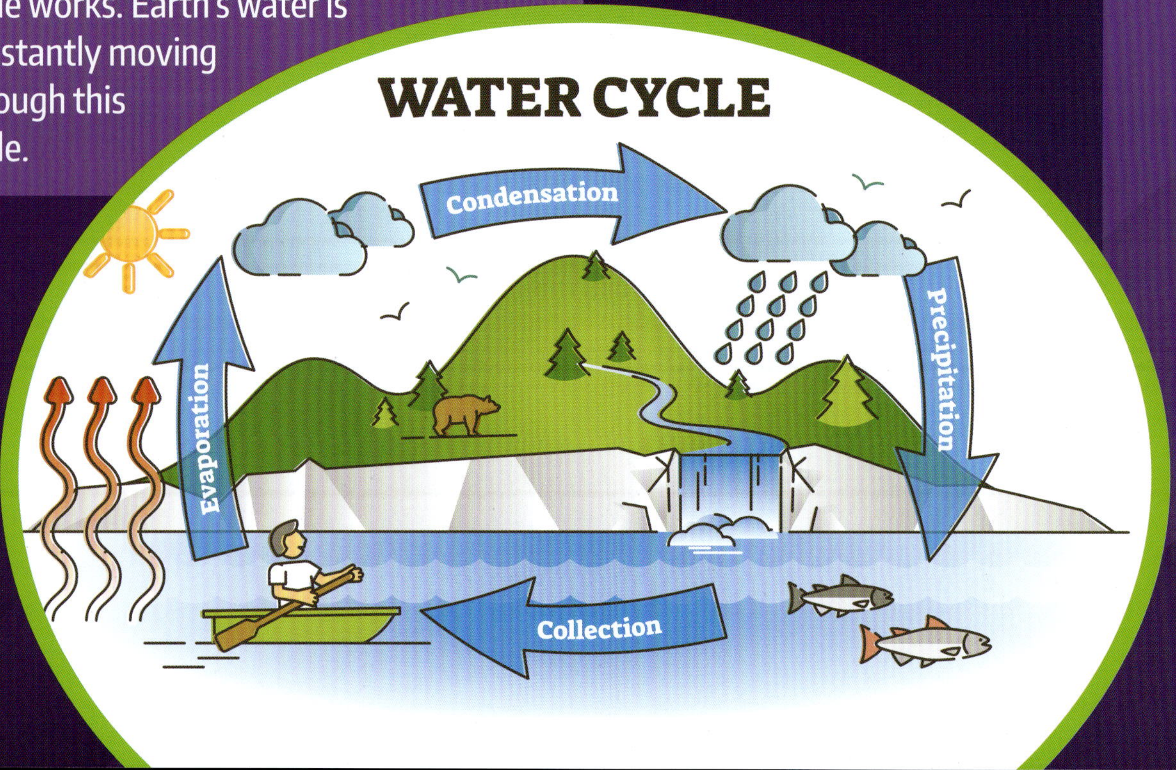

Flooding can result from heavy precipitation. Rivers become swollen and overflow. The ground also fills with water and cannot hold any more.

Rain is liquid water that falls from the sky in drops. It forms in clouds that are made up of many water droplets. When the drops become too heavy to stay in the cloud, they fall to Earth as rain. Snow is tiny ice crystals that form inside a cloud when it is cold enough. When the ice crystals become too heavy to stay in the cloud, they fall to Earth as snow.

Precipitation fills lakes, ponds, rivers, and streams. It provides the fresh water needed by humans, animals, and plants. Too little precipitation can cause drought. Too much precipitation can cause dangerous flooding.

WORD WISE

THE WATER CYCLE DESCRIBES THE WAY WATER CHANGES FORM AS IT MOVES ABOVE, WITHIN, AND AROUND EARTH.

Barometric Pressure

Atmospheric pressure, also known as barometric pressure, is the weight of air above a given area. The barometric pressure is greatest at sea level. There, the gas particles are pressed together by the weight of the air above them. Air becomes lighter farther away from Earth's surface. As height increases, the air molecules become separated by more space, and the weight decreases. As the weight of the air decreases, so does the air pressure. At sea level, air has a pressure of 14.7 pounds (6.7 kilograms) per square inch.

Barometric pressure gets lower the higher up you go.

Thunderstorms are weather events that include thick, dark clouds, heavy rain, lightning and thunder, and strong winds.

Changes in pressure help people predict approaching storms. A storm is a disturbance in the atmosphere—for example, a thunderstorm or a hurricane. When air pressure drops, it's a sign that stormy weather may be in the way. Rising air pressure is a sign of sunny weather.

Consider This

We need to breathe oxygen from the air to live. Why do you think it's harder to breathe on tall mountains than it is to breathe at sea level?

Predicting Weather

Meteorology is a field of science that studies the changes in weather on a day-to-day basis in a specific place. Scientists who study meteorology are called meteorologists.

Using various tools, meteorologists forecast whether it will rain or snow and whether it will be warm or cold. Meteorologists are often recognized as weathermen and women on television. However, most work in labs or outdoors.

Modern meteorology relies on the newest computer technology.

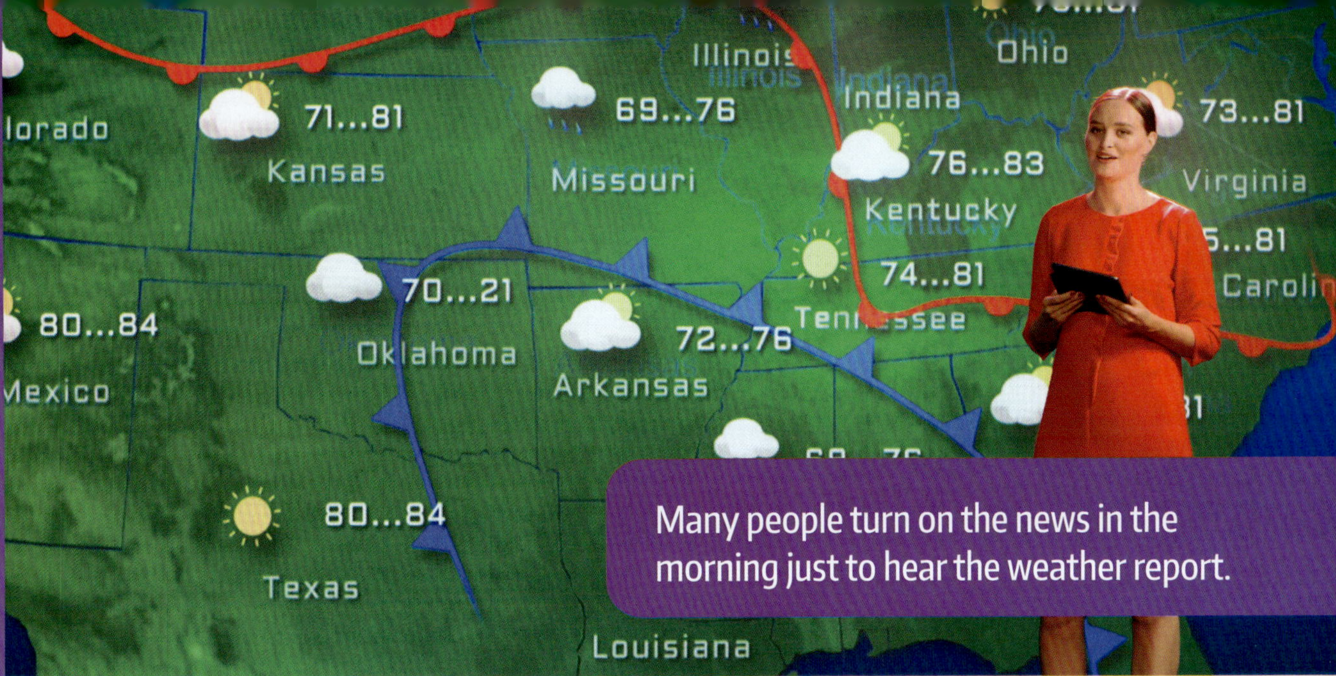

Many people turn on the news in the morning just to hear the weather report.

Meteorologists use physics and chemistry to predict weather and how it will affect Earth. These forecasts are important to many people. For example, forecasts predicting when and where hurricanes will occur allow people who might be affected to prepare for the storm. Weather forecasts aid people in the aviation, maritime, and energy industries. They also help people who work in agriculture and pollution control agencies.

Consider This

Why is it so important to so many people to know what the weather will be like on a daily basis? How can a weather report help us prepare for the day ahead?

Tools of the Trade

Long ago, people predicted the weather based on their own observations. This isn't a very accurate way of predicting weather. Today, meteorologists use many instruments to gather **data** about weather. The thermometer and the barometer are some of the oldest and most common weather instruments. The thermometer measures temperature, and the barometer measures atmospheric pressure. Another common instrument is the wind vane, which shows the direction in which the wind is blowing.

Weather stations like this one collect environmental information for use in predicting future weather events. This station can record wind speed, temperature, air pressure, and more.

The weather data that satellites collect high above Earth is sent to scientists on Earth to study and use in weather predictions.

In the twentieth century, the computer became another valuable tool in meteorology. Computer programs use mathematics to create models of weather patterns. These models can forecast the weather 10 to 30 days ahead of time. Meteorologists continue to develop and use new technology, such as radar, to study the weather on Earth.

The first weather satellite was launched in 1960. Since then, meteorologists have used satellites to photograph weather systems, such as storms, around the world. The satellites also send weather data to meteorologists.

WORD WISE

DATA ARE FACTS ABOUT SOMETHING THAT CAN BE USED IN CALCULATING, REASONING, OR PLANNING. THE SINGULAR FORM OF THE WORD IS "DATUM."

What Is Climate?

Many people think that weather and climate are the same thing, but they're not. Weather is a short-term description of the air in an area. Climate is a long-term description of the weather in a region.

The study of climates is called climatology. Scientists have many tools to help in this study. They set up weather stations around Earth to measure rainfall, temperature, and wind speed. They send weather balloons with special instruments up into the atmosphere. Weather satellites in space also report information to scientists on the ground.

Climatologists study the elements of weather over long periods. They compile climate data each month and for the year as a whole. Some climatologists study how climates change over longer periods of time, such as hundreds or thousands of years. To do this, they study historical records, but also rocks, fossils, and tree rings.

Weather agencies have long used weather balloons to study weather developments. They're still used frequently today.

compare**and**contrast

How are weather and climate the same? How are they different?

Tree rings tell us how old a tree is, but they can also reveal data about climate patterns, such as when floods and droughts occurred.

Creating Climate

Factors that influence climate include the sun, oceans, wind, land, and clouds. Sunlight affects climate by hitting Earth unevenly. Places near the equator receive lots of strong sunlight throughout the year. This gives them a hot climate year-round. Places far from the equator are cooler.

Clouds affect climate by blocking some of the heat received from the sun during the day. At night, clouds keep heat from escaping into space.

Oceans also affect climate. Land near an ocean usually has a milder climate than an inland area. The ocean warms land in winter and cools it in summer. Winds affect climate by carrying warm or cool air to areas. Wind also brings different amounts of moisture. The type of land in an area affects the climate, too. For example, mountain ranges can block cold air.

Earth's poles tilt away from the sun for half the year.
This makes them colder than most other places on Earth.

compareandcontrast

How are the effects that oceans have on climate that same or different from the effects wind has on it?

OROGRAPHIC EFFECT

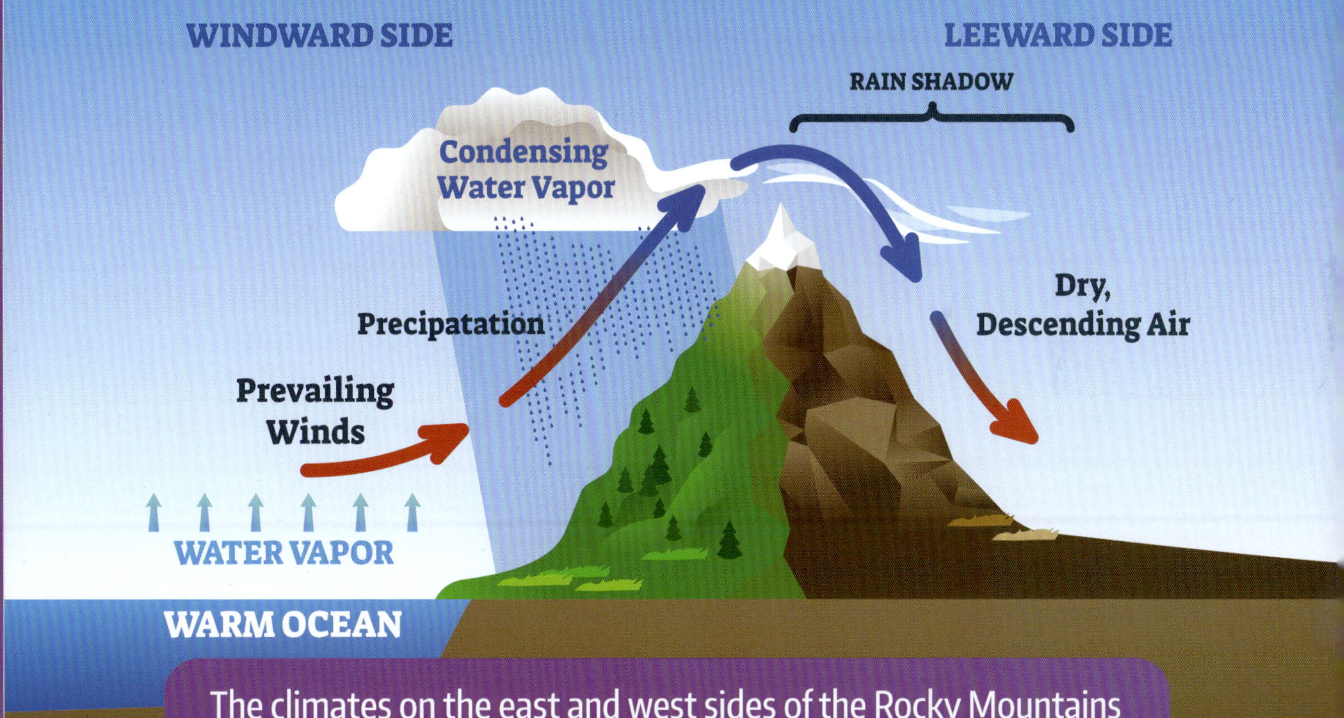

The climates on the east and west sides of the Rocky Mountains are very different. Warm air from the Pacific Ocean rises and forms rain and snow on the west side. Because of this, the east side get far less rain. This is a called a rain shadow.

Around the World

Climate differs depending on where you are. No two places on Earth have exactly the same climate. Some of the general types of climates are tropical, subtropical, cyclonic, polar, and highland.

Tropical climates are warm all year and have no winter. They lie near the equator. Some tropical climates have a lot of rain. Others are dry. Subtropical climates are found north and south of the tropical climates. They have a greater range of temperatures than tropical climates. They also may be humid or dry. Cyclonic climates are found mostly north of the equator. In these climates, people usually experience warmer summers and colder winters.

Rainforests are located in tropical climates where it is warm all year.

Polar climates are cold and snowy year-round.

Highland, or mountain, climates have a great range of temperature between day and night. They tend to be humid and cooler than the lands nearby at a lower **elevation**. Polar climates are very cold. Snow and ice often cover the land.

WORD WISE

ELEVATION IS A MEASUREMENT OF THE HEIGHT ABOVE SEA LEVEL, ALSO CALLED ALTITUDE.

What Is Climate Change?

Weather and climate change over time. This is the result of natural forces that occur over long periods of time. However, humans today are affecting Earth's climate, causing it to change more quickly.

To understand climate change, it helps to understand the greenhouse effect. A greenhouse is a glass house where plants grow. Glass lets light in and keeps heat from escaping. Gases in Earth's atmosphere, called greenhouse gases, play a similar role for the planet as a whole. They trap heat near Earth. Without the greenhouse effect, Earth would be too cold for life to exist.

For much of Earth's history, greenhouse gases were not a problem. This changed as people came to depend on fossil fuels, such as oil and coal. The burning of fossil fuels to run cars and heat homes releases large amounts of greenhouse gases into the air. Scientists think that Earth is getting warmer because of the extra greenhouse gases in the air. This is known as global warming.

Burning fossil fuels, such as coal and oil, releases greenhouse gases into the air. This is bad for the environment.

compareandcontrast

How is the greenhouse effect good for people on Earth? How is it bad for us?

The Greenhouse Effect

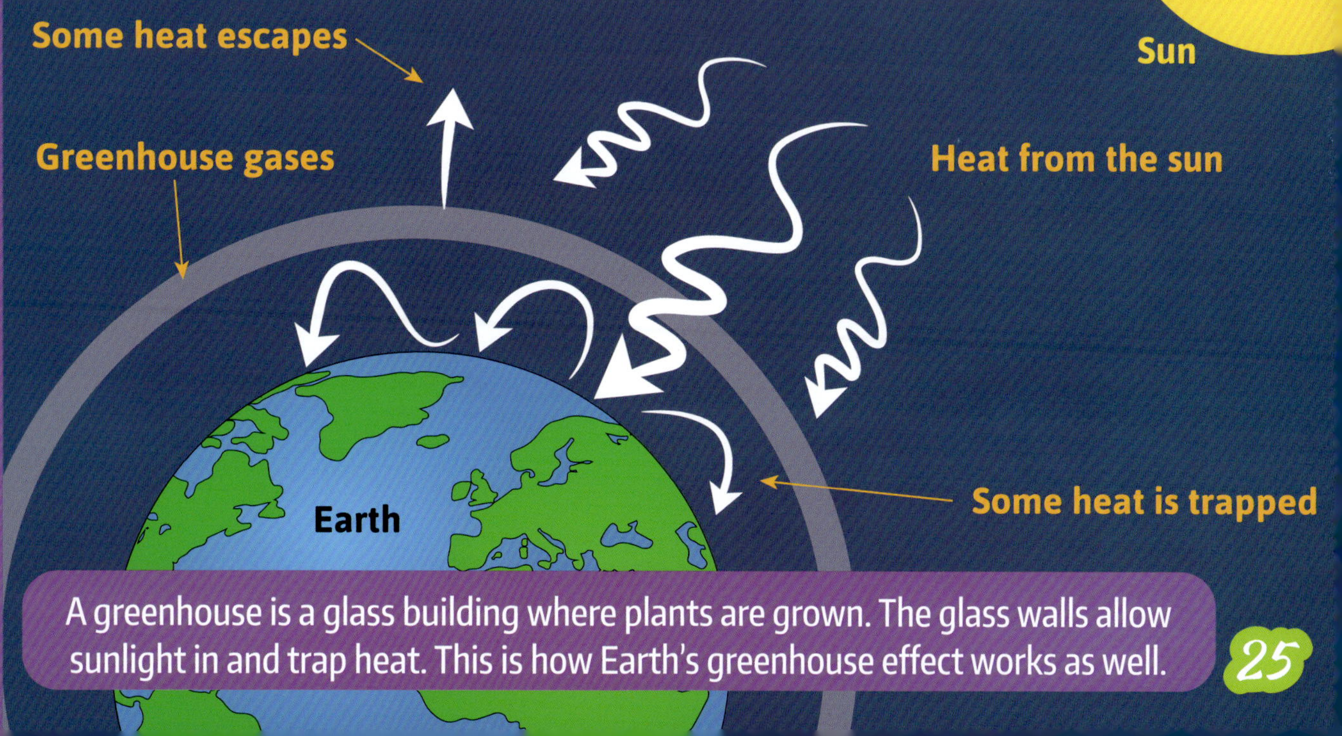

A greenhouse is a glass building where plants are grown. The glass walls allow sunlight in and trap heat. This is how Earth's greenhouse effect works as well.

Many of the activities people carry out every day help make global warming worse. For example, global warming is increased by **deforestation**. Trees absorb carbon dioxide, which is one of the greenhouse gases. Fewer trees mean that less carbon dioxide is being taken out of the atmosphere.

The effects of global warming could harm living things. The warmer temperatures could cause polar ice caps to melt. This would cause sea levels to rise. Plants, animals, and buildings along coastlines would be in danger. Many scientists believe that long-term droughts are another result of climate change.

Deforestation also destroys the homes of many animals.

People can help reduce global warming by taking public transit instead of driving cars. Some public buses run on electricity instead of fossil fuels.

Global warming is a worldwide concern. Governments are trying to limit the amount of greenhouse gases put into the air. Individual people can save energy by turning off unneeded lights and other electrical devices. This will reduce the burning of fuel that releases greenhouse gases into the atmosphere. People can also help by driving less.

WORD WISE

DEFORESTATION IS THE CLEARING, OR CUTTING DOWN, OF FORESTS, OFTEN TO MAKE MORE ROOM FOR FARMS, HOMES, AND BUSINESSES.

Climate Scholars

Throughout history, there have been many people dedicated to studying weather and climate. Without their knowledge and inventions, daily life would be more difficult.

Aristotle wrote the first study on weather in the fourth century BCE. He was a philosopher in ancient Greece who wrote about many different subjects. His *Meteorologica* was widely studied by scholars for two thousand years.

Aristotle studied weather and climate more than two thousand years ago.

Fahrenheit created a thermometer that used alcohol in 1709. In 1714, he created the first mercury thermometer.

Daniel Gabriel Fahrenheit, a German scientist, invented the modern thermometer in 1709. Today, the United States and a few other countries measure temperature with the Fahrenheit scale.

People have always observed clouds. In 1803, meteorologist Luke Howard classified and named clouds by their appearance. His research helped meteorologists make better weather predictions by connecting cloud formations to particular types of weather events.

Consider This

Which development described here do you think changed people's lives the most? Why?

Glossary

atmosphere The layer of gases that surrounds Earth.

chemistry A science that deals with the substances that make up materials and the changes that take place when substances are combined.

condense To turn from a gas to a liquid.

decrease To make or become less.

drought A long period of dry weather.

equator An imaginary circle around Earth everywhere equally distant from the North Pole and the South Pole.

formation The manner in which a thing is formed, including its structure and shape.

fossil The remains of a plant or animal that lived long ago.

fuel A material used to produce heat or power by burning.

hail Small lumps of ice that fall from clouds sometimes during thunderstorms.

molecule The smallest particle of a substance having all the characteristics of the substance.

observation An act of gathering information by noting facts or events.

philosopher Someone who seeks wisdom by thinking about important questions.

physics A science that deals with the study of physical objects and the forces (pushes or pulls) that act on them.

predict To declare in advance on the basis of study, experience, or reasoning.

research The collecting of information about a subject.

satellite A human-made object or vehicle intended to orbit Earth, the moon, or another heavenly body.

scholar A person who has done advanced study in a particular area.

For More Information

Books

Drimmer, Stephanie Warren. *National Geographic Kids Ultimate Weatherpedia*. Washington, D.C.: National Geographic Kids, 2019.

Owen, Ruth. *Let's Talk Climate Change*. Minneapolis, MN: Ruby Tuesday Books, 2026.

Websites

Daniel Gabriel Fahrenheit Facts for Kids
kids.kiddle.co/Daniel_Gabriel_Fahrenheit
Learn more about Daniel Fahrenheit, his temperature scale, and his thermometers.

Ten Basic Clouds
www.noaa.gov/jetstream/clouds/ten-basic-clouds
The National Oceanic and Atmospheric Administration explains the difference between ten basic cloud types.

Index